SIMPLE SOLUTIONS
FOR HOME BUILDING WORKERS

A BASIC GUIDE FOR PREVENTING
MANUAL MATERIAL HANDLING INJURIES

SIMPLE SOLUTIONS
TABLE OF CONTENTS

INTRODUCTION		3
1.	Soft tissue injuries	4
2.	Costs of injury	5
3.	Material handling	6
4.	Store and place materials	8
5.	Lifting and carrying	10
6.	Moving materials	12
7.	Raising and lowering	14
8.	Raise exterior walls	16
9.	Raise roof trusses	18
10.	Position and hold materials	20
11.	Repetitive handling	22
12.	Strengthen and lengthen	24
13.	Summary	26
14.	Worker protection	28
15.	Construction safety resources	30

ACKNOWLEDGEMENTS
Jason Cato (Design); Mary Ann Zapalac (Illustrations); National Council of Compensation Insurance (unpublished injury cost data, p. 5); Revised NIOSH Lifting Equation (weight limit recommendation, p. 16); Jennifer Hess, DC, PhD (pp. 24-25). Special thanks to the residential building subcontractors and workers whose participation in focus groups shaped this booklet.

DISCLAIMER
This document is in the public domain and may be freely copied or reprinted. Mention of any company or product does not constitute endorsement by the National Institute for Occupational Safety and Health (NIOSH). In addition, citations to websites external to NIOSH do not constitute NIOSH endorsement of the sponsoring organizations or their programs or products. Furthermore, NIOSH is not responsible for the content of these web sites.

SIMPLE SOLUTIONS
INTRODUCTION

Home building is physically demanding work and manual material handling may be the most difficult part of the job. Manual material handling includes all of the tasks that require you to lift, lower, push, pull, hold or carry materials.

These activities increase the risk of painful strains and sprains and more serious soft tissue injuries.

Soft tissues of the body include muscles, tendons, ligaments, discs, cartilage and nerves. Soft tissue injuries cause workers pain, suffering and lost income. They can also restrict non-work activity, like sports and hobbies. Builders' and employers' costs include loss of productivity and high workers' compensation insurance premiums.

This booklet provides basic information about readily available work practices and equipment that can help both new and experienced workers, contractors and builders prevent serious manual material handling injuries.

SOFT TISSUE INJURIES

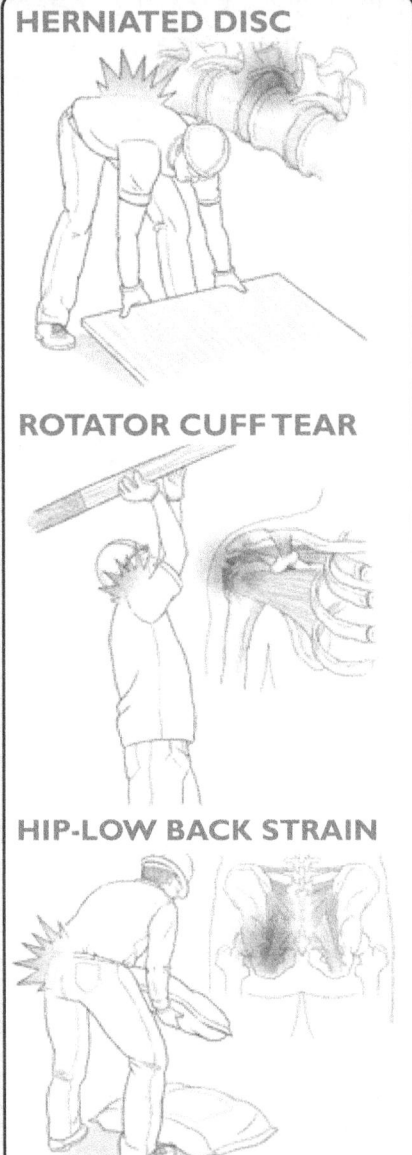

HERNIATED DISC

ROTATOR CUFF TEAR

HIP-LOW BACK STRAIN

Soft tissue injuries are different than broken bones, bruises, or punctures. They are injuries of the muscles, discs, tendons, ligaments, cartilage and nerves.

SOFT TISSUE INJURIES

- Are common with manual material handling
- Occur suddenly or develop over time
- Affect the low back, shoulders, neck, elbows, arms, hands, wrists, hips, legs, knees, ankles and feet
- Cause everyday discomfort, pain and may lead to disability
- Can take months or years to repair—if they ever do
- Interferes with work and non-work activities

COSTS OF INJURY

Workers' compensation costs for an average lost-time injury for a shoulder are $20,000 and for a back $25,000.

COSTS TO WORKERS

- Discomfort, pain & loss of income
- Restricted activities, like sports and hobbies
- Possible health care expenses

COSTS TO EMPLOYERS

- Loss of productivity
- Increased workers' compensation premiums

COSTS TO SOCIETY

- Medical expenses for uninsured workers
- Social Security disability payments

MATERIAL HANDLING
STRESS ON THE BODY

Stress on the body and risk of injury increase when you:

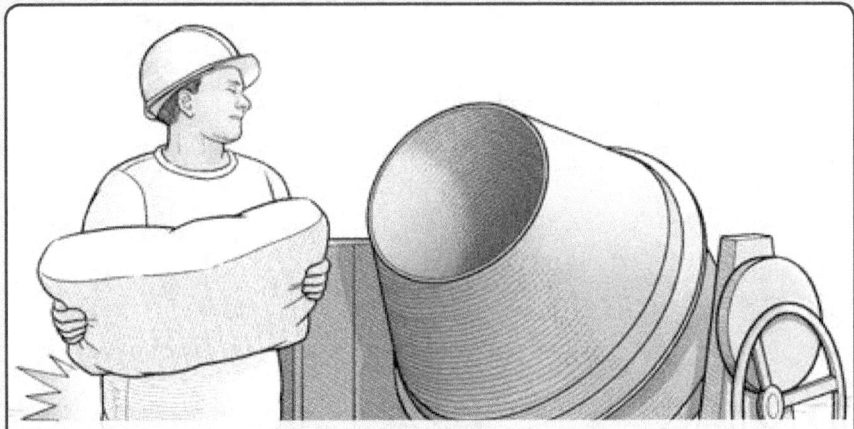

- Lift, carry, or hold heavy, unbalanced materials, especially far from the body.
- Use jerking or fast movements to lift or place materials.

- Bend and twist your back when picking up materials.

MATERIAL HANDLING
STRESS ON THE BODY

Stress on the body and risk of injury also increase when you:

- Hold materials overhead or away from the body for long periods.

- Repeatedly lift, hold, and place heavy materials.
- Hold materials away from the body.

STORE & PLACE MATERIALS
PROBLEMS

Poorly placed materials increase material handling and the risk of injury and decrease productivity.

Material storing problems commonly include:

- Not planning where materials should be staged before they are delivered.
- Staging materials far from where they will be used results in unnecessary handling.
- Storing materials too low to the ground or in confined areas makes handling more difficult. 'Waist' or 'belt' height is best.

STORE & PLACE MATERIALS
SOLUTIONS

Placing materials near the work area decreases material handling and injury risk and increases productivity.

Better material storage and placement includes:

- Plan in advance where materials will be stored when they are delivered.
- Stage materials close to where they will be used and where they will not be in other's way.
- Store materials off the ground and between knee and chest height. Leave walking space between materials.

LIFTING & CARRYING LUMBER

Avoid bending down to lift long boards of lumber. Pad your shoulder to cushion the weight of the board. Lift only one end of the board before standing and walk to the middle of the board. Rest the board on your shoulder and raise the board.

LIFTING & CARRYING
SHEET MATERIALS

It's easier to lift sheet goods off a raised pile. But when they are closer to the ground, use your legs to lift, not your back. Get as close as possible to the sheet. Raise the sheet and then tilt it so you can get a firm hold in the center. Then let the sheet become level and raise it to a comfortable position.

MOVING MATERIALS
HAND TOOLS

Using simple tools and equipment can reduce the strain on the body when carrying heavy materials.

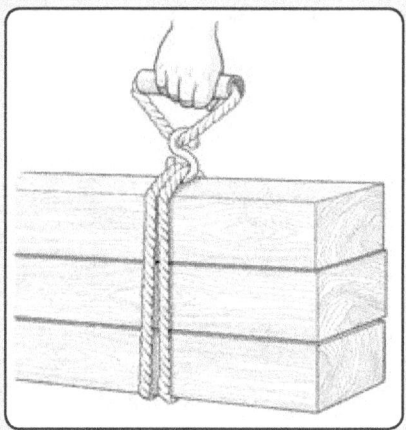

Handles made with pipe and straps for 2 people to carry heavy lumber.

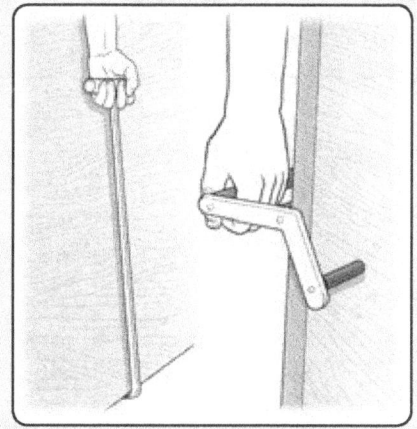

One-person & 2-person panel carriers reduce bending and make carrying easier.

Drywall dolly keeps materials off ground & moves sheet materials on floors.

Adjustable panel dolly supports heavy weights on pneumatic tires.

MOVING MATERIALS
HANDLING EQUIPMENT

Builders/contractors can provide powered equipment to move the heavier materials. Using powered equipment reduces stress on the body & increases worker productivity.

Suppliers can unload materials using truck mounted cranes or fork trucks.

Work site fork trucks can quickly move materials to where they will be used.

Skid steer loaders can move materials around on difficult building sites.

Move materials with walk-behind loaders, power-wagons and other compact equipment.

RAISING & LOWERING
PROBLEMS

Raising and lowering heavy materials increases the risk of soft tissue injuries of the back, shoulder and neck. Struck-by injuries and falls from heights can increase when heavy loads are handled between levels.

OSHA requires that one hand grasp a ladder at all times, and it prohibits holding materials while climbing up and down a ladder.

RAISING & LOWERING SOLUTIONS

Prevent injuries by handling smaller weights for less time. Use mechanical lifting equipment to eliminate unnecessary manual material handling.

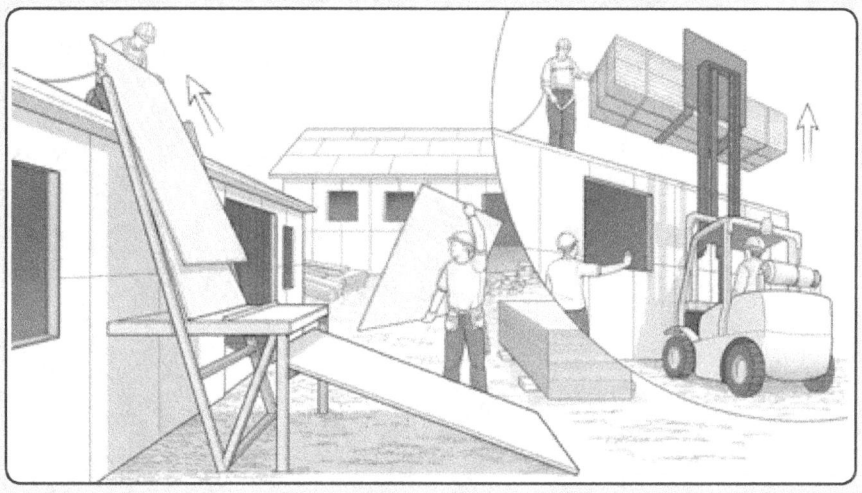

RAISE EXTERIOR WALLS
PROBLEMS

Exterior walls 12 ft. and longer weigh more than 250 lbs. Lifting walls by hand increases the risk of muscle strains and other injuries. Never lift more than 50 lbs. without getting help. Use enough people so no one lifts more than 50 lbs.

Lift with legs, not with your back. Work together, not against each other. Use wall stops and bracing to steady the wall.

RAISE EXTERIOR WALLS
SOLUTIONS

Manual and powered wall jacks can be used to raise heavy exterior walls. Smaller crews can easily raise heavy walls using the jacks, which can be purchased or rented.

RAISE ROOF TRUSSES
PROBLEMS

Trusses shorter than 20 feet can be raised by hand. Use enough people so no one lifts more than 50 lbs. Raise trusses with cranes or other equipment when possible. Balance trusses between the ropes to prevent roll.

Place & secure boards to use as a truss 'slide.' Raise trusses with rope attached to peak. Avoid bending truss.

RAISE ROOF TRUSSES
SOLUTIONS

Raise roof trusses more than 20 feet long using a crane or other equipment. Comply with OSHA fall protection and crane safety regulations, and always follow the truss manufacturer's recommendations.

POSITION & HOLD MATERIALS
PROBLEMS & SOLUTIONS

Manually holding and placing heavy steel I-beams and wood or laminated beams increases workers' risk of strains and sprains, falls, broken bones, and crushed fingers.

Placing a steel I-beam by hand across foundation wall can result in a disabling injury or death.

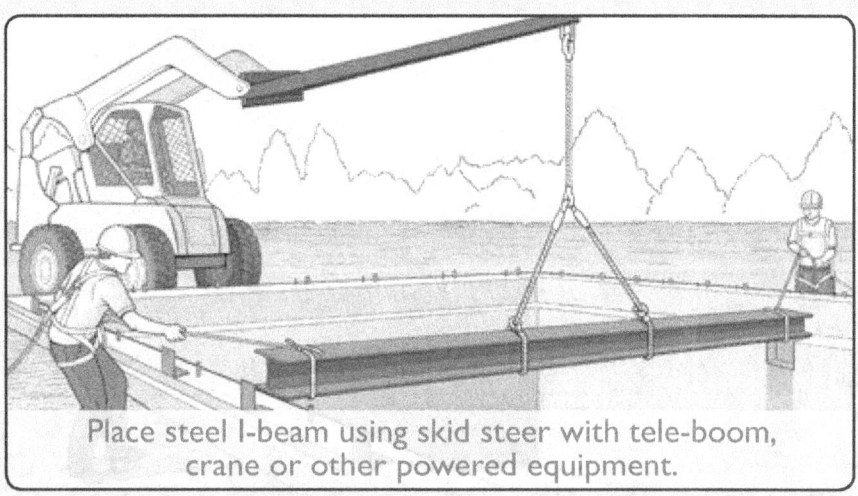

Place steel I-beam using skid steer with tele-boom, crane or other powered equipment.

POSITION & HOLD MATERIALS
PROBLEMS & SOLUTIONS

Manually holding and positioning sheet materials overhead, like drywall, can strain the neck, back, shoulder and arm muscles. Use tools like T-braces or mechanical lifts to hold sheets against the ceiling.

Handling drywall overhead fatigues the muscles and can lead to neck and shoulder injuries.

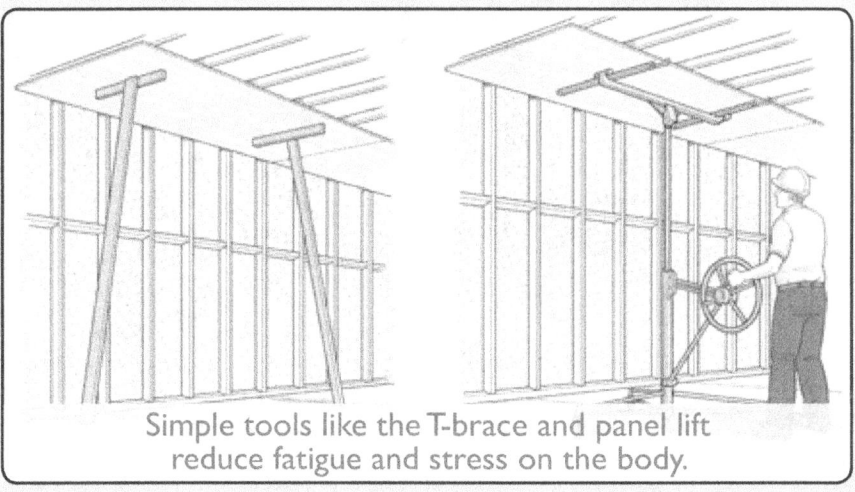

Simple tools like the T-brace and panel lift reduce fatigue and stress on the body.

REPETITIVE HANDLING
PROBLEMS

Repeatedly handling heavy blocks and other materials puts extra stress on your body. The weight of the materials and awkward body positions – like frequent bending, reaching, and twisting – increase your chance of a muscle or joint injury.

REPETITIVE HANDLING
SOLUTIONS

Change the way you do the work to reduce injury risks. Place materials close to where they are needed. Set up the work to reduce bending and twisting. Keep materials close to your body. Take short breaks to give muscles and joints needed rest-time.

Increase height of blocks to reduce bending & twisting.

Use two-level veneer or mason scaffold for higher walls.

STRENGTHEN & LENGTHEN
MUSCLE EXERCISES

Improve core strength and muscle tone with these exercises before work or during breaks. Exercise slowly, don't bounce!

Slowly make large circles forward and backward with each arm while marching in place. Continue for 1 min.

Stand upright with arms relaxed. Step forward slowly with one foot. Do not move knee past your ankle. Keep trunk erect. Return to standing. Continue for 1 min. Switch legs & repeat.

Stand upright with arms relaxed. Take a large step to the left then return to standing. Repeat by stepping to right. Continue for 1 min.

Stand upright with arms relaxed. Take 5 side-steps to the right. Take 5 side-steps to the left. Repeat 5 times.

STRENGTHEN & LENGTHEN
MUSCLE EXERCISES

If you have an existing muscle, joint or disc injury, or experience pain with exercise, consult your doctor before doing exercises.

Hold bar (or pretend to) behind neck, arms bent at elbows 90°. Gently pull bar backward away from head until you feel stretch in the front of shoulders. Hold 12 seconds, relax. Perform 5 times.

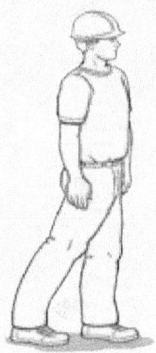

Stand straight, extend one leg backward, contracting buttock muscles. Keep trunk upright. Hold 10 sec, perform 3 times, each side.

Stand straight, lock stomach muscles by pulling rib cage and pelvis together as shown. Try to mildly tighten stomach muscles (10%) when lifting objects. Hold 12 sec, perform 10 times.

Place hands on hips as pictured. Slowly bend backwards, keeping knees straight. Do not extend your head. Hold 12 sec, perform 5 times.

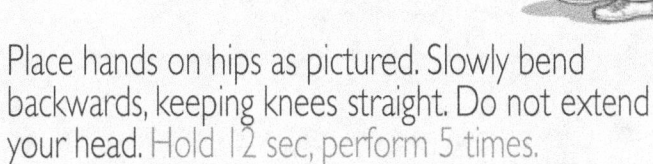

SUMMARY
PREVENTING INJURIES

MANUAL MATERIAL HANDLING can cause strains & sprains and more serious injuries to your body. These injuries often result in pain, time away from work and lost wages. Even less serious injuries can keep you from enjoying non-work activities, like sports and hobbies.

You can reduce your chance of serious injury by using safe work practices and following the recommendations below:

Staging materials far from where they will be used and close to the ground increases injury risks.

- Plan ahead to save time and effort.
- Decide in advance where YOU want the materials placed when they're delivered.
- Keep materials off the ground to reduce stressful bending and lifting.

Bending and twisting your body when lifting heavy materials increases the risk of muscle and other soft tissue injuries.

- Don't lift and carry more than 50 lbs. alone. Get help from coworkers.
- Bend your knees and push up with your legs.
- Hold materials close to your body.
- Lift heavier lumber at one end – not the center – and walk to the center to lift it.
- Use tools and equipment to transport heavy materials when possible.

SUMMARY
PREVENTING INJURIES

Raising and lowering heavy materials to different work levels increases the risk of soft tissue and other serious injuries.

- Lift, hold, and carry materials close to your body.
- Use supports and equipment to hold materials overhead.
- Use platforms for raising materials to different work heights.
- NEVER carry materials in your hands on ladders.
- NEVER lift or position heavy materials standing on a ladder.
- Use mechanical equipment to raise and lower heavier materials.
- Use fall protection as required when working at heights and raising or lowering materials.

Holding unsupported materials above the shoulders fatigues the shoulders and neck and can result in serious injury.

- Use tools and equipment to support materials.
- NEVER support heavy materials on your head.
- Take short breaks to give muscles and joints time to 'recover' from the strain.
- Use tools & equipment to support heavy loads and reduce your strain.

Repeatedly lifting and positioning heavy materials – like concrete blocks – increases the physical stress on the same muscles and soft tissues.

- Use boards or scaffolds to keep blocks, mortar, and other materials around knee high.
- Don't twist the body when lifting or placing materials.

WORKER PROTECTION
RIGHTS & RESPONSIBILITIES

Employers must insure their employees have a workplace free of recognized job hazards that can cause serious injury or death. Federal and State Occupational Safety and Health Administrations (OSHA) enforce job safety and health regulations to protect workers.

WORKERS' SAFETY & HEALTH RIGHTS INCLUDE:

- Taking action alone or with co-workers to protect your safety and health.
- Contacting OSHA to request a safety inspection of your job site.

EMPLOYERS' SAFETY RESPONSIBILITIES INCLUDE:

- Informing employees about job hazards through training and other means.
- Training employees in a language and vocabulary they can understand.
- Providing certain types of personal protective equipment (PPE), including fall protection.

If you live in one of the States (or Puerto Rico) shown below, you can get your State OSHA contact info by calling Federal OSHA (1-800-321-6742) or visiting http://www.osha.gov/dcsp/osp/index.html.

Alaska, Arizona, California, Hawaii, Indiana, Iowa, Kentucky, Maryland, Michigan, Minnesota, Nevada, New Mexico, North Carolina, Oregon, South Carolina, Tennessee, Utah, Vermont, Virginia, Washington, Wyoming, and Puerto Rico.

If you live in other States or US Territories, contact Federal OSHA at: (Tel) 1-800-321-6742 or find the contact information for the nearest Federal OSHA Regional or Area office by visiting http://www.osha.gov/html/RAmap.html

WORKER PROTECTION
RIGHTS & RESPONSIBILITIES

Employers sometimes classify workers as "independent contractors", rather than as employees. Employees have legal rights to minimum wage, overtime pay, Workers' Compensation, job site safety and health and filing **OSHA** complaints. "Independent contractors" do not have these protections.

For more information call 866–487–9243 or visit http://www.dol.gov/whd/workers/misclassification/

WORKERS' COMPENSATION INSURANCE

- Employers must have Workers' Compensation insurance to pay employees' injury-related medical costs and other benefits in every state except Texas. Without Workers' Compensation benefits workers may not receive the medical care or other benefits they deserve.

- For info regarding individual State Workers' Compensation programs, visit http://www.dol.gov/owcp/dfec/regs/compliance/wc.htm

FEDERAL AND STATE WAGE LAWS

- Federal and State laws require that employers pay employees a minimum wage for the regular hours they work. If you work more than 8 hours per day or 40 hours per week, you may be eligible for a higher wage for the extra hours you work. For info regarding Federal wage and overtime pay requirements, call 1–866–487–2365 or visit http://www.dol.gov/whd/.

- For info regarding individual State wage and overtime pay requirements, visit http://www.dol.gov/whd/contacts/state_of.htm

CONSTRUCTION SAFETY RESOURCES

For more information about preventing work related injuries and illnesses, you can check out the information provided by the following organizations:

OSHA'S RESIDENTIAL CONSTRUCTION REGULATIONS
Description of Occupational Safety & Health Administration (OSHA) safety & health regulations.
http://www.osha.gov/SLTC/residential/index.html

NIOSH CONSTRUCTION INFORMATION
Free information about safety and health hazards in the construction industry.
http://www.cdc.gov/niosh/construction/

CENTER FOR CONSTRUCTION RESEARCH & TRAINING
Source for information about controlling and eliminatingconstruction safety and health hazards and training.
http://www.cpwrconstructionsolutions.org/

WASHINGTON STATE RESIDENTIAL CONSTRUCTION SAFETY INFORMATION
http://www.lni.wa.gov/safety/topics/atoz/topic.asp?KWID=252

NATIONAL ASSOCIATION OF HOME BUILDERS
Safety & health information from home builders' trade association.
http://www.nahb.org/page.aspx/category/sectionID=616

CALIFORNIA FRAMING CONTRACTORS ASSOCIATION
Source for safety information.
http://www.californiaframingcontractors.org/

INTERFAITH WORKER JUSTICE
Affiliated Worker Centers provide safety & health training in English and Spanish and assist workers with other employment problems, like "wage theft".
http://www.iwj.org/network/workers-centers

CONSTRUCTION SAFETY RESOURCES

Lifting and carrying more than 50 lbs. increases your risk of low back injury. Use the list below to help keep the weight you handle to around 50 lbs.

LUMBER PIECES (KILN DRIED)

4	10 ft.	2"x4"	51 lbs.
3	12 ft.	2"x4"	46 lbs.
2	10 ft.	2"x6"	40 lbs.
2	12 ft.	2"x6"	48 lbs.
2	10 ft.	2"x8"	53 lbs.
1	10 ft.	2"x10"	66 lbs.
1	10 ft.	2"x12"	41 lbs.
2	10 ft.	4"x4"	60 lbs.

LVL (LAMINATED VENEER LUMBER) PIECES

1	10 ft.	1 3/4"x9 1/4"	47 lbs.
1	10 ft.	1 3/4"x11 7/8"	61 lbs.
1	10 ft.	1 3/4"x14"	71 lbs.

SHEETS— PLYWOOD / OSB

2	3/8 in.	4'x8'	68 / 77 lbs.
1	1/2 in.	4'x8'	45 / 54 lbs.
1	5/8 in.	4'x8'	58 / 67 lbs.
1	3/4 in.	4'x8'	68 / 80 lbs.

SHEETS — CEMENTITIOUS BACKBOARD

1	7/16 in.	4'x8'	96 lbs.

CONCRETE BLOCKS — LIGHT WT. / NORMAL WT.

1	6"x8"x16"		22 / 34 lbs.
1	8"x8"x16"		27 / 44 lbs.
1	12"x8"x16"		35 / 55 lbs.

www.ingramcontent.com/pod-product-compliance
Lightning Source LLC
Chambersburg PA
CBHW070732180526
45167CB00004B/1725